TARANTULAS

Cody Koala

An Imprint of Pop!
popbooksonline.com

abdobooks.com

Published by Pop!, a division of ABDO, PO Box 398166, Minneapolis, Minnesota 55439. Copyright © 2022 by Abdo Consulting Group, Inc. International copyrights reserved in all countries. No part of this book may be reproduced in any form without written permission from the publisher. Cody Koala™ is a trademark and logo of Pop.

Printed in the United States of America, North Mankato, Minnesota.

052021
092021

THIS BOOK CONTAINS RECYCLED MATERIALS

Cover Photo: Shutterstock Images
Interior Photos: Shutterstock Images, 1, 5 (bottom left), 5 (bottom right), 9, 10, 13, 20; iStockphoto, 5 (top), 6–7; Phil Degginger/Science Source, 15 (top), 16; Scott Linstead/Science Source, 15 (bottom left); Robert Shantz/Alamy, 15 (bottom right); R Kawka/Alamy, 19

Editor: Aubrey Zalewski
Series Designers: Laura Graphenteen and Colleen McLaren

Library of Congress Control Number: 2020948905
Publisher's Cataloging-in-Publication Data
Names: Golkar, Golriz, author.
Title: Tarantulas / by Golriz Golkar
Description: Minneapolis, Minnesota : Pop!, 2022 | Series: Desert animals | Includes online resources and index.
Identifiers: ISBN 9781532169755 (lib. bdg.) | ISBN 9781098240684 (ebook)
Subjects: LCSH: Tarantulas--Juvenile literature. | Spiders--Juvenile literature. | Arachnids--Juvenile literature. | Desert animals--Juvenile literature.
Classification: DDC 591.754--dc23

Hello! My name is

Cody Koala

Pop open this book and you'll find QR codes like this one, loaded with information, so you can learn even more!

Scan this code* and others like it while you read, or visit the website below to make this book pop.

popbooksonline.com/tarantulas

*Scanning QR codes requires a web-enabled smart device with a QR code reader app and a camera.

Table of Contents

Chapter 1

Hairy Spiders

Tarantulas are large, hairy spiders. There are more than 850 kinds of tarantulas in the world. Many live in the desert. They live alone.

Watch a video here!

hairs
eyes
jaws
legs

Desert tarantulas are mostly brown or black. They have eight eyes and eight legs. They have strong jaws and sharp **fangs**.

Chapter 2

Living in a Burrow

Many desert tarantulas live in Mexico and the southwestern United States. They dig **burrows** to hide from **predators**. Burrows also keep tarantulas cool.

Complete an
activity here!

burrow with silk lining

Tarantulas line their
burrows with spider silk.
The silk keeps dirt out.
Tarantulas may leave lines
of silk outside their burrows.
The silk shakes when
touched. This tells a tarantula
that **prey** or a predator
is nearby.

If a predator comes
too close, a tarantula attacks
it. The tarantula scrapes
needlelike hairs off its
stomach with its back legs.
Then it flings the hairs at the
predator to scare it away.

Male tarantulas live up to 12 years.
Females live up to 25 years.

Chapter 3

Hunting at Night

Tarantulas hunt for food at night. They seek out their **prey**. Tarantulas eat mostly insects. Sometimes they eat lizards, snakes, or mice.

Learn more here!

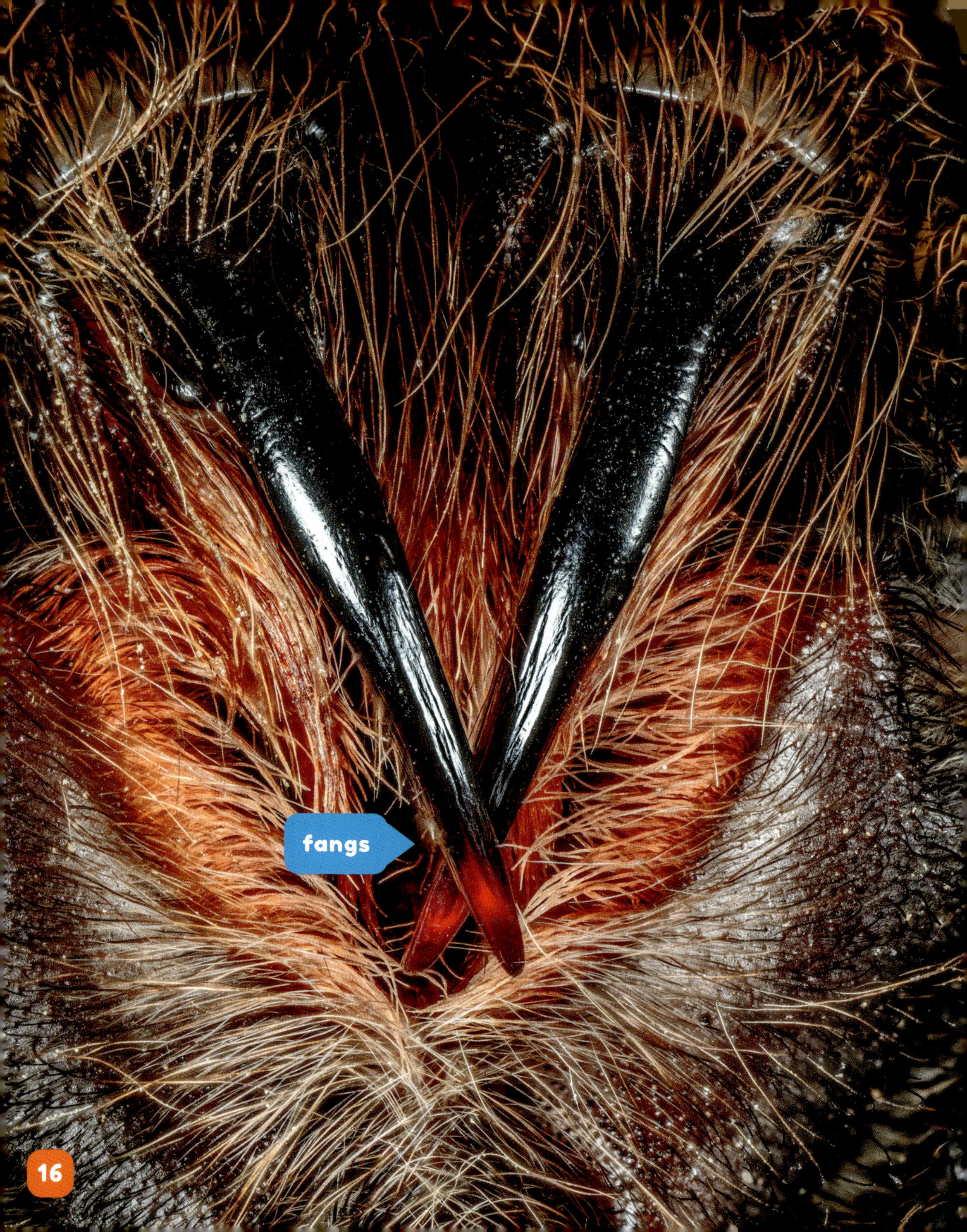

fangs

Tarantulas jump onto their prey. They grab prey with their legs. They bite down. **Venom** from their **fangs** turns prey into liquid. This makes prey easier to swallow.

Chapter 4

Young Tarantulas

Desert tarantulas **mate** in summer and fall. The female lays 500 to 1,000 eggs inside an egg sac. The babies hatch after six to nine weeks.

Learn more here!

tarantula molting

Baby tarantulas live on their own right after they hatch. They **molt** to grow. When they molt, they grow new outer skeletons. They lie on their backs. Then they push off their old skeletons.

Tarantulas can regrow their stomach linings and lost legs.

Making Connections

Text-to-Self

Tarantulas keep cool by resting in their burrows. What do you like to do when the weather is hot?

Text-to-Text

Have you read a book about another animal that has venom? How is that animal like a tarantula? How is it different?

Text-to-World

Mother tarantulas make egg sacs to protect their eggs. How do other animals protect their babies?

Glossary

burrow – a hole that an animal digs in the ground for shelter.

fang – a spider's mouthpart that has venom at the tip.

mate – to come together to have babies.

molt – to shed an outer layer.

predator – an animal that hunts other animals for food.

prey – an animal that is hunted by other animals.

venom – poison from an animal bite or sting.

Index

babies, 18, 21

burrows, 8, 10, 11

fangs, 7, 16, 17

hairs, 4, 6, 12

molting, 20, 21

prey, 11, 14, 17

spider silk, 10, 11

venom, 17

Online Resources

popbooksonline.com

Thanks for reading this Cody Koala book!

Scan this code* and others like it in this book, or visit the website below to make this book pop!

popbooksonline.com/tarantulas

*Scanning QR codes requires a web-enabled smart device with a QR code reader app and a camera.